Table of Con

INTRODUCTIO 7

CHAPTER 1 13

The Story of Cider 13

Ancient Origins......................................14

Medieval Europe: A Drink of the People 15

Colonial America and Farmhouse Traditions...18

Decline and Industrialization..............20

The Modern Craft Revival....................21

Cider Today: A Global Tradition.........23

Why History Matters for the Home Cidermaker...24

Conclusion..25

CHAPTER 2 27

Apples and Their Characteristics..................27

Understanding Apple Categories........28

The Chemistry of Flavor......................31

Choosing Apples for Cider 33

Heritage vs. Modern Apples 34

Foraging, Growing, and Sourcing Apples 37

Blending: The Art of Balance 39

Conclusion 40

CHAPTER 3 42

Equipment and Preparation 42

Basic Equipment for Beginners 43

Intermediate and Advanced Equipment 47

Sanitation: The Cornerstone of Success 49

DIY and Budget-Friendly Alternatives 51

Preparing Your Workspace 53

Safety Considerations 54

Conclusion 56

CHAPTER 4 58

Equipment and Tools for Cidermaking 58

1. Basic Equipment for Beginners 59

2. Optional But Helpful Additions 65

3. Scaling Up: Advanced Tools 67

4. Choosing the Right Setup for You 70

5. Caring for Your Equipment.......................71

CHAPTER 5 73

Harvesting and Pressing Apples...................73

1. Knowing When Apples Are Ready...........74

2. Choosing Apples for Cider........................76

3. Preparing Apples for Pressing.................78

4. Crushing the Apples 79

5. Pressing the Apples.................................81

6. Collecting and Protecting Juice 83

7. Estimating Yield.......................................85

8. From Orchard to Fermenter 86

CHAPTER 6 87

Fermentation Fundamentals 87

1. What Is Fermentation? 88

2. Wild Fermentation vs. Cultured Yeast 90

3. Preparing the Juice..................................92

4. The Fermentation Vessel 94

5. The Fermentation Process Step by Step ..95

6. Temperature Control 97

7. Signs of Healthy Fermentation................99

8. Common Challenges 100

9. Deciding When It's Done........................101
CHAPTER 7 103
Blending and Flavor Development.............103
1. Why Blending Matters 104
2. The Key Elements of Balance 106
3. Timing Your Blend.................................108
4. Techniques for Blending 109
5. Methods of Flavor Development 111
6. Common Blending Strategies.................113
7. Practical Tips for Blending Success 114
8. The Creative Side of Blending 116
CHAPTER 8 117
Bottling, Carbonation, and Storage 117
The Importance of Bottling Done Right 118
Choosing Bottles and Closures...................119
Glass Bottles.......................................119
Closures 121
Carbonation: Still or Sparkling?.................122
Still Cider..122
Sparkling Cider...................................123
Calculating Priming Sugar 125

The Bottling Process: Step by Step 127
Storage and Aging 128
Safety Considerations 130
Bringing It All Together 130
CHAPTER 9 132
Troubleshooting and Fine-Tuning 132
Common Problems and Their Causes 133
1. Stuck Fermentation 133
2. Cloudy Cider 135
3. Unwanted Sweetness or Dryness 136
4. Off-Flavors 138
5. Bottle Bombs 139
Prevention Is Better Than Cure 141
Fine-Tuning Your Craft 142
Experiment with Yeast Strains 142
Blending for Balance 143
Adjusting Acidity and Tannins 144
Aging Techniques 144
When to Start Over 145
The Cidermaker's Mindset 146
THE END 147

INTRODUCTION

Cider is more than a beverage—it is a story in a glass. From the apple orchards of ancient Europe to the farmhouse presses of colonial America, cider has been at the heart of communities, celebrations, and everyday life for centuries. Long before beer and wine became staples, apples were being transformed into a drink that was refreshing, nourishing, and deeply tied to the land. Today, with the rise of artisanal food and drink, cider has returned to the spotlight, combining tradition with modern creativity.

This book invites you into that world. Whether you are a beginner curious about your first batch or an enthusiast looking to refine your craft, you will find both guidance and inspiration here. By the time you reach the final chapter, you will not only understand the process of turning apples into cider but also develop the confidence to make ciders that express your own style.

We begin by exploring the **history and heritage of cider**—a journey that stretches from the orchards of medieval England, France, and Spain to the craft cider houses of today. This foundation provides context for the practical steps that follow.

You will then learn about **apples themselves**, the soul of cider. Not all apples are created equal, and understanding their varieties, flavors, and qualities is the first step in creating a balanced drink. From there, we move into the **tools and equipment** you need, showing how to get started with simple setups and how to upgrade as your skills grow.

The heart of the book guides you through the cidermaking process:

- **Harvesting and pressing apples into juice.**
- **Fermentation fundamentals**—wild yeasts, cultured yeasts, timing, and monitoring.
- **Blending and flavor development**, where artistry meets science.
- **Maturation and aging**, the stage where cider deepens and evolves.
- **Bottling and carbonation**, including how to safely produce still or sparkling cider.

No craft comes without challenges, so you will also find a chapter dedicated to **troubleshooting and fine-tuning**. Here, we'll cover how to identify common issues, prevent mistakes, and rescue a batch if things don't go as planned.

Finally, the book closes with **recipes, pairings, and serving ideas**—from classic dry cider to mulled seasonal blends. You'll also discover suggestions for pairing cider with food, helping you present it not just as a drink but as part of a larger dining experience.

Cidermaking is both science and art, but above all, it is a craft that connects you to nature and tradition. With this book, you hold a guide that combines historical insight, practical instruction, and creative inspiration. Whether you dream of enjoying a crisp glass of homemade cider at the end of the day or hope to share your creations with friends and family, your journey begins here—one apple, one press, and one sip at a time.

CHAPTER 1

The Story of Cider

Cider's history is as rich and varied as the drink itself. From its ancient beginnings to today's craft revival, cider has accompanied human societies through agriculture, trade, and culture. Understanding its story not only deepens our appreciation for a glass of cider but also places our own cidermaking efforts within a much larger tradition.

Ancient Origins

The earliest evidence of cider-like beverages dates back thousands of years. While the exact origins are hard to pinpoint, apples have grown wild across Europe and Asia since prehistoric times, and fermentation was almost certainly discovered by accident. People likely collected fallen apples, stored them, and found that their juice naturally transformed into a pleasantly intoxicating drink.

By the time of the Roman Empire, cider was already established in several parts of Europe. Julius Caesar himself noted the popularity of apple-based drinks when he invaded the British Isles in 55 BCE. The Romans were skilled at cultivating orchards, and they spread improved apple varieties across their territories, laying the groundwork for cider traditions that still exist today.

Medieval Europe: A Drink of the People

Cider truly flourished during the Middle Ages. In France's Normandy and Brittany regions, cider became a staple drink, rivaling wine in popularity. The cool, damp climate of northern France was better suited to apple orchards than grape vineyards, making cider the natural choice for farmers and peasants alike. French cidermakers developed the bittersweet and bittersharp apples that continue to define the region's ciders today.

Across the English Channel, cider was equally important. By the 14th century, cider was being produced on nearly every farm in England's cider-making counties such as Herefordshire, Somerset, and Devon. It was consumed daily by both the working class and the gentry. In many cases, cider was safer to drink than water, which was often contaminated. Employers even paid farm laborers in cider as part of their wages.

Spain, too, developed its own cider culture, particularly in the northern region of Asturias. Unlike the sweet and rounded ciders of Normandy, Asturian cider was sharp, dry, and effervescent, poured dramatically from a height to aerate the drink and release its aroma. This ritual pouring, or *escanciado*, remains a proud cultural tradition today.

Colonial America and Farmhouse Traditions

When European settlers arrived in North America, they brought apple seeds and grafts with them. The native crabapple was too small and sour for cider, so settlers cultivated orchards of European apple varieties. By the 17th and 18th centuries, cider had become the most common household beverage in colonial America.

Beer was difficult to produce due to limited access to barley and hops, but apples thrived in the new world's climate. Every homestead typically had an orchard, and nearly every family made its own cider. Founding fathers such as John Adams and Thomas Jefferson were known to be cider enthusiasts, and cider was consumed daily at meals, often in place of water.

Cider also became an integral part of American community life. Local presses were busy each autumn, and cider was stored in barrels for the year ahead. Variations included "ciderkin" (a weaker drink made from re-pressing apple pomace with water) and stronger distilled apple spirits like applejack. Cider, in its many forms, was the drink of the people.

Decline and Industrialization

The 19th and early 20th centuries marked a turning point for cider. In Europe, industrialization introduced large-scale beer brewing, which quickly overtook cider in popularity due to lower production costs and consistent supply. In America, the arrival of German immigrants brought beer culture to the forefront, while the rise of grain farming further reduced reliance on orchards.

Prohibition in the United States (1920–1933) nearly destroyed cider culture altogether. Orchards were cut down, and cidermakers were forced to abandon their craft. Even after Prohibition ended, beer and spirits dominated, leaving cider largely forgotten except in small pockets of traditional production in Europe.

The Modern Craft Revival

The late 20th and early 21st centuries witnessed cider's remarkable comeback. In the United Kingdom, small-scale producers revived regional cider traditions, emphasizing natural fermentation and heritage apple varieties. In France and Spain, cider never fully disappeared, but renewed interest in artisanal food helped elevate it to new heights.

In the United States, a new wave of cidermakers emerged in the 1980s and 1990s, fueled by the broader craft beer and farm-to-table movements. Today, cider is one of the fastest-growing sectors of the beverage industry. Craft cidermakers experiment with both tradition and innovation—producing everything from bone-dry farmhouse ciders to hopped, spiced, or fruit-infused variations.

Cider Today: A Global Tradition

What makes cider unique today is its diversity. Each cider-producing region brings its own identity to the drink:

- Normandy's bittersweet balance.
- England's bold, tannic farmhouse styles.
- Spain's sharp, effervescent sidra.
- America's experimental blends and creative flavors.

This diversity is a strength, offering something for every

palate. Whether enjoyed at a rustic table in Asturias, a festival in Somerset, or a tasting room in Oregon, cider connects us to centuries of tradition while still offering room for modern creativity.

Why History Matters for the Home Cidermaker

You may wonder why the history of cider is so important when your main goal is to make a good batch at home. The answer is simple: cidermaking is both an art and a heritage. By learning how people across centuries and continents crafted their cider, you gain insights into methods, styles, and flavors that can inspire your own approach.

Every time you press an apple or watch yeast transform juice into cider, you are participating in a tradition that stretches back thousands of years. That awareness not only deepens your appreciation but also gives your work a sense of belonging in a story much larger than yourself.

Conclusion

Cider's story is one of resilience, adaptation, and rediscovery. It has endured wars, prohibition, industrialization, and shifting tastes. Yet it continues to thrive, renewed by each generation of makers and drinkers.

As you begin your own cidermaking journey, remember that you are carrying this legacy forward. From ancient orchards to modern glasses, cider has always been a drink that connects people to land, culture, and community. In the chapters ahead, you will learn how to transform apples into cider step by step—but always with an appreciation for the history that brought us here.

CHAPTER 2

Apples and Their Characteristics

If cider has a soul, it is the apple. Just as wine depends on the grape, cider depends entirely on the apple's unique balance of sugar, acidity, and tannins. Choosing the right apples is the first and most important decision a cidermaker makes. This chapter explores the different types of apples, the chemistry that defines their flavors, and the art of blending them into a cider that is greater than the sum of its parts.

Understanding Apple Categories

Not all apples are created equal when it comes to cider. Dessert apples—the ones you find in grocery stores—are often too sweet and lack the tannins or acidity necessary for a balanced cider. Traditional cider apples fall into four broad categories:

1. **Sweet Apples** – High in sugar, low in acidity and tannins. They provide the fermentable sugar that yeast transforms into alcohol. While they lack sharpness,

they add body and sweetness to a blend.

    - Examples: Golden Delicious, Fuji, Gala.

2. **Sharp Apples** – High in acidity, low in tannins. These apples bring brightness and freshness, balancing sweetness with a crisp bite.
    - Examples: Granny Smith, Bramley, Winesap.
3. **Bittersweet Apples** – High in tannins, moderate sugar, low acidity. These are prized in traditional English and French ciders for their

ability to give structure, complexity, and astringency.

    - Examples: Dabinett, Yarlington Mill, Muscadet de Dieppe.

4. **Bittersharp Apples** – High in tannins and acidity, often too intense to drink as juice on their own but invaluable in small amounts for adding depth.
    - Examples: Kingston Black, Foxwhelp.

By understanding these categories, cidermakers can design blends that achieve balance: sweetness for alcohol, acidity for freshness, and tannins for structure.

The Chemistry of Flavor

Apples may appear simple, but their composition is complex. Three main elements determine how cider will taste:

- **Sugar** – The fuel for fermentation. More sugar means higher alcohol content. Sweet apples or late-harvest fruit generally provide the richest sugar content.

- **Acidity** – Gives cider its sharpness and balance. Too little acidity produces a flat drink; too much can make it sour and unpalatable.
- **Tannins** – Compounds that create bitterness and astringency, providing body and a drying sensation on the palate. Without tannins, cider can feel thin or one-dimensional.

Beyond these, other compounds—such as esters, phenols, and malic acid—contribute aroma, flavor, and mouthfeel. The cidermaker's role is to understand how these natural elements interact during fermentation and aging.

Choosing Apples for Cider

The ideal apples for cider depend on what style you want to make. A light, refreshing cider might focus on sweet and sharp apples, while a traditional farmhouse cider will lean heavily on bittersweet and bittersharp varieties.

- **For balanced ciders:** Mix two to three categories (e.g., 50% sweet, 30% sharp, 20% bittersweet).
- **For dry, tannic ciders:** Use a base of bittersweet

apples, with just enough sharp fruit for acidity.

- **For bright, refreshing ciders:** Rely on sharp apples, softened with sweet varieties for alcohol content.

One of the joys of cidermaking is experimentation. Even small changes in apple proportions can dramatically shift the flavor of the final product.

Heritage vs. Modern Apples

Supermarket apples are bred for shelf life, appearance, and crisp texture—qualities that don't necessarily translate into good cider. Traditional cider apples, on the other hand, are often unattractive, bitter, or even inedible when eaten raw, yet they make extraordinary cider.

- **Heritage Cider Apples** – Varieties like Kingston Black, Dabinett, and Foxwhelp have been cultivated for centuries specifically for cider. They are rare outside of orchards dedicated to cidermaking but worth seeking out for authenticity and depth.

- **Modern Apples** – Varieties like Gala or Fuji are more accessible, and while they may lack tannins, they can still produce pleasant ciders, especially when blended with sharper fruit.

If heritage apples are not available, mixing modern dessert apples with wild or crab apples (which are naturally sharp and tannic) can create surprisingly good results.

Foraging, Growing, and Sourcing Apples

How you obtain apples will shape your cidermaking journey:

- **Foraging** – Many cidermakers begin by foraging from abandoned orchards, hedgerows, or friends' backyards. These apples may be irregular, but they often contain unique flavors missing from commercial varieties.
- **Growing Your Own** – Planting a small orchard or grafting cider apple varieties ensures a steady supply, but

requires patience and knowledge of soil, climate, and pruning.

- **Local Orchards and Markets** – Farmers' markets and small orchards may sell mixed apples ideal for cider. Some orchards even press juice to order.
- **Commercial Sources** – If you are unable to gather your own fruit, commercial juice blends can still be fermented into decent cider, though with less complexity.

Blending: The Art of Balance

The secret to great cider lies not in any single apple but in the harmony of many. Just as winemakers blend grapes, cidermakers blend apples to balance sugar, acidity, and tannins.

- A cider made from only sweet apples may be alcoholic but bland.
- A cider from only sharp apples may be overly sour.
- A cider from only bitter apples may be harsh and astringent.

But when blended thoughtfully, the result can be rich, refreshing, and full of character. Blending is part science—guided by measurements of sugar and acidity—and part intuition, honed over years of tasting and adjusting.

Conclusion

The apple is the foundation of every cider. By learning to identify apple categories, understand their chemistry, and blend them effectively, you set yourself up for success as a cidermaker. While equipment, yeast, and technique all play vital roles, nothing influences the final drink as much as the fruit you start with.

As you explore orchards, markets, or even your own backyard, begin to taste apples not just as fruit but as potential cider. Notice their sweetness, sharpness, and bitterness. Ask yourself how they might combine with others. In doing so, you will begin to see apples in a new light—not simply as fruit, but as the building blocks of a craft that connects you to centuries of tradition and to the glass of cider you are about to create.

CHAPTER 3

Equipment and Preparation

Cidermaking has been practiced for centuries with tools ranging from rustic wooden presses to today's stainless-steel fermenters. While equipment has evolved, the principles remain the same: extract juice from apples, allow yeast to ferment it, and carefully guide the process to produce a drink that is safe, stable, and delicious.

This chapter covers the essential tools needed for cidermaking, options for upgrading as you grow more serious, and the critical importance of cleanliness and preparation. Whether you are working in a home kitchen or a backyard shed, the right equipment—and how you care for it—will make all the difference.

Basic Equipment for Beginners

You don't need a professional setup to make good cider. With a few simple tools, you can start producing small batches right at home.

- **Apple Crusher (or Chopper):** Apples must be

broken down before pressing. For small batches, you can chop apples by hand or use a food processor. For larger quantities, a hand-crank apple crusher makes the work faster and more consistent.

- **Cider Press:** This is used to squeeze juice from crushed apples. Traditional wooden presses are iconic, but modern plastic or stainless-steel models are easier to clean. For beginners, even a clean kitchen strainer and muslin cloth can serve as a makeshift press.

- **Fermentation Vessel (Demijohn or Carboy):** Once you have juice, it needs to ferment in a sealed container. Glass demijohns, food-grade plastic buckets, or stainless-steel fermenters all work. Choose a vessel that allows you to see the liquid and is large enough to hold your intended batch.
- **Airlock:** An airlock fits onto the fermentation vessel to release carbon dioxide while keeping out oxygen and contaminants. This small but crucial device prevents spoilage.

- **Siphon Tube:** Used for transferring cider from one vessel to another without disturbing sediment.
- **Hydrometer:** A simple tool to measure sugar levels and track fermentation progress. It helps you estimate alcohol content and know when fermentation is complete.

With just these items, you can successfully make cider at home.

Intermediate and Advanced Equipment

As your passion grows, you may want to invest in better tools for consistency and quality.

- **Refractometer:** More precise than a hydrometer for measuring sugar levels, especially in small samples.
- **pH Meter or Strips:** Measures acidity in your juice, allowing for fine adjustments.
- **Temperature Control:** Yeast is sensitive to temperature. A temperature-controlled fermentation

chamber, heating belt, or cooling system helps maintain stability.

- **Stainless Steel Fermenters:** Durable, easy to clean, and resistant to scratches that harbor bacteria.
- **Filtering and Clarifying Tools:** For those seeking crystal-clear cider, filtration units or fining agents can be used.

While these tools are not necessary for beginners, they can greatly improve control over the process as you refine your skills.

Sanitation: The Cornerstone of Success

If there is one principle that separates good cider from spoiled cider, it is sanitation. Apples themselves carry wild yeasts, bacteria, and natural molds. While many of these organisms contribute positively to fermentation, others can ruin a batch. Proper cleaning ensures that only the yeasts you want take control.

- **Clean Before Use:** Wash all equipment with hot water and scrub away visible residue.

- **Sanitize Before Use:** Apply a no-rinse sanitizer (such as potassium metabisulfite solution or food-grade sanitizing agents) to all vessels, tubing, and tools that come into contact with juice.
- **Repeat at Every Stage:** Each time cider is transferred, racked, or bottled, re-sanitize equipment to prevent contamination.

Cidermakers often say: *you cannot over-sanitize*. Neglecting this step is one of the most common mistakes among beginners.

DIY and Budget-Friendly Alternatives

If you are hesitant to invest in equipment right away, there are many creative solutions.

- **Crushing Apples:** A rolling pin, food processor, or even freezing and thawing apples can break them down enough for pressing.
- **Pressing Juice:** A clean cloth bag and heavy weight can act as a press for small

batches. Some people use a potato ricer or juicer.

- **Fermentation Vessels:** Food-grade plastic buckets with tight lids (often sold for brewing beer) are inexpensive and widely available.
- **Airlocks:** In a pinch, a balloon with a pinhole can substitute, though a proper airlock is more reliable.

These methods may lack efficiency, but they prove that cidermaking is accessible to anyone with patience and creativity.

Preparing Your Workspace

Before starting a batch, take time to prepare both your equipment and environment.

- **Choose a Stable Location:** Fermentation vessels should be stored in a cool, dark place with a stable temperature. Avoid direct sunlight and wide fluctuations in heat or cold.
- **Organize Tools in Advance:** Gather everything you need before crushing apples or transferring juice. Cider

moves quickly once pressing begins.

- **Keep Records:** Many cidermakers maintain a brewing log with details such as apple varieties used, sugar levels, fermentation temperature, and tasting notes. Over time, this record becomes invaluable for refining recipes.

Safety Considerations

While cidermaking is generally safe, a few precautions are worth noting:

- **Glass Carboys:** Always handle glass with care, especially when full. Use a carrier or protective net to prevent accidents.
- **Presses and Crushers:** Keep hands and clothing away from moving parts.
- **Sanitizers and Chemicals:** Use food-safe sanitizers only, and handle them according to instructions.

By respecting the process and your tools, cidermaking can be an enjoyable and safe craft for years to come.

Conclusion

Equipment is the backbone of cidermaking. From simple presses and airlocks to advanced fermenters and testing instruments, the right tools make the journey smoother and the results more consistent. Yet, the true key lies not in how expensive your setup is but in how carefully you prepare and care for it.

A beginner with basic tools and excellent sanitation can produce a cider every bit as enjoyable as a professional working with stainless steel tanks. The difference is mindfulness: cleaning thoroughly, handling apples with respect, and understanding what each piece of equipment contributes to the process.

With your tools ready, sanitized, and organized, you are now prepared for the next step—turning ripe apples into fresh juice, the foundation of every great cider. That journey begins in the orchard and carries us to the press, which we will explore in the next chapter.

CHAPTER 4

Equipment and Tools for Cidermaking

Making cider is an age-old craft, and while the heart of the process lies in the apples and the fermentation, the tools you use will greatly influence your experience and results. The good news is that cidermaking does not require overly complicated or expensive equipment. With a few essential tools, you can make small batches at home; and as your skills grow, you can gradually expand your toolkit for larger, more refined production.

This chapter walks you through the equipment needed, from the simplest setups for beginners to more advanced options for those looking to refine their craft. By the end, you'll know not only what to use but also why each tool matters.

1. Basic Equipment for Beginners

If you're just starting out, you don't need to invest heavily. Many of the tools required are either inexpensive or may already be in your kitchen. Here's a starter list:

- **Apple Crusher or Chopper**

Before pressing, apples must be broken down into a pulp. For small batches, a sturdy knife or food processor will do. For larger batches, a hand-cranked apple crusher speeds things up considerably, breaking the fruit into manageable pieces that release juice more easily.

- **Apple Press** This is the centerpiece of cidermaking equipment. Traditional basket presses or modern ratchet presses are common. They use pressure to squeeze juice from the apple pulp. For the hobbyist, a small tabletop press is usually enough. If you want a budget-friendly option, even a clean kitchen strainer bag and hand-squeezing can work for very small batches.

- **Fermentation Vessel** Once you have juice, it needs a home for fermentation. Glass carboys (large glass jugs) or food-grade plastic buckets with tight-sealing lids are the most common choices. Carboys are excellent for monitoring fermentation since they're transparent, but buckets allow easier handling and cleaning.

- **Airlock and Bung** Fermentation produces carbon dioxide gas, which must escape without letting oxygen or contaminants in. This is where the airlock comes in. It's a simple, inexpensive device filled with water that allows CO_2 to bubble out while protecting your cider. The bung (stopper) fits the carboy or bucket opening and holds the airlock securely.

- **Hydrometer** This essential tool measures the sugar content of your apple juice, helping you estimate potential alcohol content and track fermentation progress. By taking readings before and after fermentation, you can calculate the strength of your cider with accuracy.
- **Siphon Tube** When moving cider from one vessel to another (for racking, bottling, or aging), a siphon tube lets you do it gently, leaving sediment

behind and minimizing oxygen exposure.

- **Bottles and Caps/Corks** Once fermentation is complete, you'll need containers to store your cider. Swing-top bottles, capped beer bottles, or corked wine bottles all work. Choose based on whether you want sparkling or still cider. A hand capper or corker will be necessary for sealing.

2. Optional But Helpful Additions

As you become more comfortable, you may find that investing in a few additional tools will improve efficiency and quality:

- **pH Strips or Meter** These allow you to measure acidity levels, which is crucial for flavor balance and stability. Proper acidity helps prevent spoilage and ensures a crisp, refreshing cider.
- **Thermometer** Fermentation temperature greatly affects yeast activity

and cider flavor. A thermometer helps you monitor conditions, especially if your brewing space fluctuates in temperature.

- **Sanitizing Solutions** Cleanliness is key. While you can use boiling water for some items, commercial no-rinse sanitizers (like Star San) make it easy to keep equipment sterile and prevent contamination.
- **Secondary Carboys** Having multiple fermentation vessels lets you rack your cider (transfer it

off the sediment) into a clean container, promoting clarity and avoiding off-flavors from yeast breakdown.

3. Scaling Up: Advanced Tools

If you plan to produce cider in larger quantities or experiment with professional-level techniques, consider the following:

- **Electric Apple Crusher** For bigger harvests, a motorized crusher saves significant time and effort, producing consistent pulp for pressing.

- **Hydraulic or Rack-and-Cloth Press**
  These larger presses can handle bigger volumes and extract juice more efficiently, making them ideal for community cider projects or small commercial ventures.
- **Stainless Steel Fermenters**
  While glass and plastic work well, stainless steel fermenters offer durability, ease of cleaning, and the ability to control temperature more precisely.
- **Temperature-Controlled Fermentation Chambers**

Consistency is vital for refining cider flavor. A fermentation fridge or chamber allows you to set and maintain the ideal temperature for yeast activity.

- **Filtration Systems** To achieve crystal-clear cider, some cidermakers use filtration setups to remove yeast and particulates before bottling. This is optional but can elevate presentation.

4. Choosing the Right Setup for You

The best equipment for cidermaking depends on your goals:

- If you want a **simple, rustic experience**, a knife, kitchen juicer, and a fermentation bucket may be all you need.
- If you're aiming for **consistent, high-quality batches**, investing in a press, hydrometer, and secondary vessels will serve you well.

- If you're **dreaming of semi-professional production**, stainless steel fermenters and temperature control will help you fine-tune flavor and stability.

Remember: cidermaking is flexible. Start with what you have, add tools as you grow, and let your cider guide your investments.

5. Caring for Your Equipment

Whatever setup you choose, cleanliness is non-negotiable. Dirty equipment is the number one cause of failed batches. Always:

- **Clean thoroughly** after each use to remove pulp, juice, and residue.
- **Sanitize** before using equipment again to eliminate unwanted bacteria or wild yeasts.
- **Store equipment properly** in a dry, safe space to extend its lifespan.

Good equipment care not only protects your cider but also ensures consistent results year after year.

CHAPTER 5

Harvesting and Pressing Apples

The journey from orchard to cider glass begins with one of the most rewarding stages of the process—harvesting and pressing apples. This stage is both practical and symbolic: it's where you connect directly with the fruit, select the flavors that will define your cider, and experience the satisfaction of transforming apples into golden juice ready for fermentation.

In this chapter, we'll explore when and how to harvest apples, how to prepare them, and the different methods of pressing to ensure you capture every drop of character they have to offer.

1. Knowing When Apples Are Ready

Not all apples ripen at the same time, and timing is everything. Apples harvested too early can be starchy and lack flavor; too late, and they may ferment on their own before you can use them. The key is understanding ripeness.

- **Color and Skin:** Most apple varieties change color as they ripen, developing richer tones. Look for vibrant hues rather than pale, dull skins.

- **Taste Test:** The simplest method—bite into one. A ripe cider apple should taste flavorful, with balanced sweetness and acidity. Bitter or astringent cider apples often mellow during fermentation, but they should still have depth.
- **Ease of Picking:** Ripe apples detach easily when lifted and twisted gently from the branch. If they resist, they're not ready.

- **Falling Fruit:** When apples begin dropping naturally, it's a good sign harvest is near. However, avoid using bruised or rotting windfall apples, as they can introduce spoilage.

2. Choosing Apples for Cider

The blend of apples you harvest will shape the character of your cider. As discussed earlier, apples fall into four categories: sweet, sharp, bittersharp, and bittersweet. Combining them thoughtfully ensures balance:

- **Sweet apples** provide sugar (fuel for fermentation).
- **Sharp apples** add refreshing acidity.
- **Bittersweet and bittersharp apples** contribute tannins and complexity.

For small batches, don't worry if you can't get the "perfect" mix—simply experiment. Even supermarket apples can yield pleasant cider, though traditional cider apples bring more depth.

3. Preparing Apples for Pressing

Before pressing, apples should be prepared carefully to ensure clean, flavorful juice.

- **Wash Thoroughly:** Rinse apples in clean water to remove dirt, debris, and potential contaminants. For larger harvests, a bucket of water or hose spray works well.
- **Sort and Inspect:** Remove any apples that are badly bruised, moldy, or rotten. A little blemish is fine, but rot can spoil a whole batch.

- **Resting Apples:** Some cidermakers "sweat" apples—storing them in a cool, dry place for a week or two after harvest. This allows starches to convert into sugars and flavors to concentrate. Sweated apples are softer and press more easily, producing richer juice.

4. Crushing the Apples

Whole apples cannot be pressed effectively; they need to be broken down into a pulp, or "pomace," first. This step increases juice yield and ensures full extraction.

- **Manual Crushing:** For small batches, you can chop apples with a knife or use a food processor.
- **Hand-Crank Crusher:** A traditional apple mill or crusher uses rotating blades or rollers to shred apples into uniform pulp. This is the most common method for hobbyists.
- **Electric Crusher:** For larger quantities, electric or motorized crushers save time and produce consistent pomace.

The goal is a fine but not mushy pulp—pieces small enough to release juice but not so fine that they clog your press.

5. Pressing the Apples

Pressing transforms apple pulp into juice, and the method you choose depends on your scale and budget.

- **Hand-Squeezing:** For very small batches, you can place pulp in a clean muslin or strainer bag and squeeze by hand. This is labor-intensive and yields limited juice but works in a pinch.

- **Basket Press:** The most traditional type, with a wooden or metal basket where pulp is placed and pressure is applied via a screw or ratchet. Simple, effective, and ideal for small to medium batches.
- **Rack and Cloth Press:** Pulp is spread between cloths stacked on racks, then pressed. This method produces high yields and is used by many traditional cideries.
- **Hydraulic Press:** A more advanced option, using hydraulic pressure for

efficiency. Excellent for larger batches and commercial use.

No matter the method, aim to extract as much juice as possible without crushing seeds, which can impart bitterness.

6. Collecting and Protecting Juice

Freshly pressed apple juice is called "must," and it's the lifeblood of your cider. At this stage, it's highly perishable and must be handled carefully.

- **Use Food-Grade Containers:** Collect juice in sanitized buckets or carboys.
- **Avoid Oxygen Exposure:** Pour gently to minimize splashing, which can lead to oxidation.
- **Strain as Needed:** Filtering the juice through cheesecloth or a mesh strainer removes pulp and debris for a cleaner fermentation.

Some cidermakers choose to pasteurize juice by heating or add Campden tablets (potassium metabisulfite) to kill wild yeasts and bacteria before pitching their chosen yeast. Others embrace wild fermentation, letting natural yeasts on the apples drive the process. The choice depends on whether you prefer control or tradition.

7. Estimating Yield

How much juice can you expect? On average:

- **10–15 pounds of apples** yield about **1 gallon (4.5 liters)** of juice.
- Yield depends on apple variety, ripeness, and pressing method.

Keep this in mind when planning your harvest and batch size.

8. From Orchard to Fermenter

Once you have your juice, it's time to move quickly into fermentation. Transfer the juice into sanitized vessels, fit with an airlock, and allow the transformation to begin. This transition marks the shift from hands-on orchard work to the quieter, scientific magic of fermentation, which we'll explore in the next chapter.

CHAPTER 6

Fermentation Fundamentals

If pressing apples is the most physical part of cidermaking, fermentation is the most magical. At this stage, golden apple juice begins its transformation into cider through the action of yeast. These tiny organisms consume the natural sugars in the juice and convert them into alcohol, carbon dioxide, and a host of flavor compounds that give cider its depth and character.

For beginners, fermentation can seem intimidating—a mysterious process that happens out of sight in sealed vessels. But in truth, it is both straightforward and deeply rewarding. With the right knowledge and a little attention, you can guide fermentation to produce consistently delicious results. This chapter will cover everything you need to know: the science behind fermentation, choosing yeast, managing conditions, and recognizing when your cider is ready to move on.

1. What Is Fermentation?

Fermentation is the natural metabolic process where yeast consumes sugars and releases alcohol and carbon dioxide. In cider, the sugar comes primarily from fructose and glucose found in apples.

The reaction is simple: **Sugar → Alcohol + Carbon Dioxide + Flavor Compounds**

While simple in principle, the process is influenced by many factors—temperature, yeast strain, oxygen, nutrients, and acidity. Each variable can subtly shift the character of the finished cider. Understanding these fundamentals allows you to control and refine your product.

2. Wild Fermentation vs. Cultured Yeast

One of the first choices you'll face is whether to ferment your juice with wild yeast or by adding a cultured yeast strain.

- **Wild Fermentation** Apples naturally carry wild yeasts on their skins, and if juice is left alone, these will start fermentation on their own. The advantage is authenticity and complexity—wild fermentations can create unique, layered flavors. The drawback is unpredictability; sometimes wild yeasts stall or produce off-flavors.

- **Cultured Yeast** Commercial cider and wine yeasts are carefully selected strains that provide consistency, clean fermentation, and control over flavor. Different strains bring different qualities—some accentuate fruitiness, others create drier profiles. For most beginners, cultured yeast is the safer and more reliable option.

Many cidermakers experiment with both approaches—using cultured yeast for their main batches while reserving a portion of juice for wild fermentation to explore complexity.

3. Preparing the Juice

Before adding yeast, check a few key qualities of your juice:

- **Sugar Content (Gravity):** Use a hydrometer to measure specific gravity. This tells you how much sugar is present and helps estimate potential alcohol. Typical apple juice ranges from 1.045 to 1.065, which produces cider around 5–8% ABV.

- **Acidity (pH):** Ideal cider pH is around 3.3–3.8. Too high (low acidity), and cider may taste flat and risk spoilage; too low (high acidity), and it may taste harsh.
- **Nutrients:** Apple juice is often low in yeast nutrients. Adding a small amount of yeast nutrient can ensure a healthy fermentation, especially for larger batches.

Some cidermakers also add Campden tablets at this stage to kill wild yeasts and bacteria, then wait 24 hours before pitching cultured yeast. This ensures only your chosen yeast controls fermentation.

4. The Fermentation Vessel

Your vessel is the yeast's home for several weeks or months, so it must be suitable:

- **Glass Carboys:** Transparent, easy to monitor, and excellent for small batches.

- **Food-Grade Buckets:** Affordable and easy to handle; best for larger volumes.
- **Stainless Steel Fermenters:** Durable and professional, offering precise control.

Whatever vessel you choose, always sanitize thoroughly before use. Contamination is the most common cause of spoiled cider.

5. The Fermentation Process Step by Step

1. **Pitching the Yeast:** Add yeast to your prepared juice.

If using dried yeast, rehydrate it according to the manufacturer's instructions.

2. **Primary Fermentation:** Over the first few days, yeast multiplies rapidly. You'll see bubbling in the airlock and froth forming on the surface. This vigorous phase may last a week.
3. **Secondary Fermentation:** After the initial burst, activity slows. Yeast continues consuming sugars more steadily, clarifying the cider as sediment (lees) settles at the bottom.

4. **Racking:** To avoid off-flavors, cider is often siphoned (racked) into a clean vessel after a few weeks, leaving sediment behind. This can be repeated as needed.
5. **Completion:** Fermentation is complete when the cider's gravity remains stable over several days, usually between 0.998 and 1.010 depending on dryness.

6. Temperature Control

Temperature has a powerful effect on yeast behavior:

- **Cool Fermentation (10–15°C / 50–59°F):** Produces clean, crisp flavors with subtle aromatics. Often preferred for traditional ciders.
- **Moderate Fermentation (16–20°C / 60–68°F):** Balanced speed and flavor expression.
- **Warm Fermentation (21–25°C / 70–77°F):** Faster, fruitier fermentations, but risk of off-flavors increases.

Consistency is key. Fluctuations can stress yeast, leading to stalled fermentation or undesired flavors. If possible, ferment in a space where temperature is stable.

7. Signs of Healthy Fermentation

- Steady bubbling in the airlock.
- A yeasty, fruity aroma without unpleasant sulfur or rotten smells.
- Gradual clearing of cider as yeast settles.

If activity slows too quickly, yeast nutrient or gentle agitation may help. Always check gravity with a hydrometer before deciding fermentation is stuck.

8. Common Challenges

- **Stuck Fermentation:** Yeast stops working before all sugars are consumed. Causes include low nutrients, cold temperatures, or stressed yeast. Solutions: warm the cider slightly, add nutrient, or pitch fresh yeast.
- **Off-Flavors:** Sulfur smells, vinegar notes, or excessive sweetness can indicate issues. Often, proper sanitation and temperature

control prevent these problems.

- **Overflows:** In vigorous fermentations, foam can push through the airlock. Leaving headspace or using a blow-off tube helps.

9. Deciding When It's Done

Fermentation is complete when:

- Gravity readings are stable for at least three days.
- Airlock activity slows to a stop.
- The cider tastes dry or balanced, depending on your recipe.

At this point, you can either move to aging for complexity or bottle immediately for fresh, youthful cider.

CHAPTER 7

Blending and Flavor Development

While fermentation turns apple juice into cider, blending and flavor development are what elevate it from a simple drink into an art form. Every batch of cider reflects the fruit, yeast, and process used, but blending allows you to fine-tune balance, correct imperfections, and create unique styles that reflect your vision as a cidermaker.

In traditional cider regions like England, France, and Spain, blending has long been considered the cidermaker's craft. Rarely is a cider made from a single apple variety; instead, multiple juices and styles are brought together, each contributing different qualities—sweetness, acidity, tannins, body, or aroma. This chapter will guide you through the principles of blending, techniques for enhancing flavor, and ways to experiment creatively without losing balance.

1. Why Blending Matters

Apples vary widely in character. Some are high in sugar but lack acidity; others are sharp but thin;

still others are tannic and complex yet bitter on their own. By blending, you combine strengths and offset weaknesses. The result is a cider that feels complete: refreshing yet full-bodied, smooth yet lively.

Blending also gives you:

- **Consistency:** Apples vary year to year. Blending allows you to maintain a house style despite natural fluctuations.
- **Complexity:** Different apples (and even different fermentations) contribute layers of aroma and taste.

- **Flexibility:** If one batch ferments too dry, too sharp, or too tannic, blending can bring it into harmony.

2. The Key Elements of Balance

When developing flavor, consider the balance between these main elements:

- **Sweetness:** Adds body and approachability. Comes from residual sugars or back-sweetening.
- **Acidity:** Provides freshness and crispness. Sharp apples

or malic acid adjustments contribute here.

- **Tannins:** Add structure, dryness, and astringency. Common in bittersweet and bittersharp apples.
- **Aroma and Flavor:** Fruity esters from yeast, subtle wild notes, or spiced/mulled additions.

Think of cider like music—sweetness is the melody, acidity the rhythm, tannins the bass, and aroma the color. A good blend brings them into harmony.

3. Timing Your Blend

Blending can happen at several stages, each with advantages:

- **Pre-Fermentation:** Combine juices from different apple varieties before fermentation. This creates integrated flavors from the start and simplifies the process.
- **Post-Fermentation (Before Aging):** Blend finished ciders once fermentation is complete. This allows you to assess each cider's character and combine them strategically.

- **Before Bottling:** Final tweaks can be made just before bottling. This is common for adjusting sweetness, carbonation, or aroma.

Many cidermakers use a combination—pre-fermentation blending for overall style and post-fermentation blending for fine adjustments.

4. Techniques for Blending

- **Small-Scale Trials:** Before committing to a full batch, blend samples in small glasses or test tubes. Experiment with different ratios until you find a profile you like.

- **Gradual Adjustments:** Start with a base cider and add measured amounts of another, tasting as you go. Keep notes so you can scale up later.
- **Blending Across Vintages:** If you plan to make cider over multiple years, blending older and younger ciders can add depth and consistency.

Remember: once blended, flavors may need time to "marry." What tastes sharp today may smooth out in a few weeks.

5. Methods of Flavor Development

Beyond blending apple varieties or finished ciders, you can enhance flavor through other techniques:

- **Back-Sweetening:** Adding sugar, honey, or juice after fermentation to increase sweetness. Must be stabilized (with Campden and sorbate) if you don't want fermentation to restart.

- **Infusions:** Adding spices (cinnamon, cloves, ginger), herbs (rosemary, mint), or even botanicals like hops during or after fermentation.
- **Fruit Additions:** Blending in juice or puree from pears, berries, or stone fruits for unique styles.
- **Oak Aging:** Resting cider in oak barrels (or with oak chips) adds vanilla, caramel, or smoky notes.
- **Malolactic Fermentation:** A secondary fermentation where malic acid converts to lactic acid, softening

sharpness and adding creamy, buttery undertones.

Each technique should complement, not overpower, the natural apple character.

6. Common Blending Strategies

- **Dry and Sweet Blend:** Mix a very dry cider with a sweeter one to achieve semi-dry balance.
- **Sharp and Tannic Blend:** Combine a high-acid cider with a tannic one to create structure without harshness.

- **Aromatic Lift:** Add a small portion of a fruity, estery cider to boost aroma.
- **Wild and Clean Fusion:** Blend a wild-fermented cider with a cultured one—capturing rustic notes while maintaining control.

7. Practical Tips for Blending Success

- **Keep Records:** Document ratios and tasting notes. This helps replicate or refine blends in future batches.

- **Use Neutral Bases:** A clean, balanced cider can serve as a canvas for more distinctive batches.
- **Taste in Context:** Evaluate cider at the temperature and glassware in which it will be served; cold temperatures can mute sweetness and aroma.
- **Don't Overcomplicate:** Sometimes two or three well-chosen components are better than a dozen.

8. The Creative Side of Blending

Blending is not only about fixing or balancing—it's an opportunity to express creativity. Some cidermakers craft seasonal blends, such as spiced winter ciders or berry-infused summer editions. Others create "signature styles" that reflect their orchards, much like winemakers with estate wines.

Trust your palate, but also be open to feedback from friends, family, or tasting groups. What excites you may surprise others, and cider is best enjoyed in community.

CHAPTER 8

Bottling, Carbonation, and Storage

For many cidermakers, bottling is the most exciting stage of the journey. After weeks—or sometimes months—of fermenting, blending, and aging, your cider is ready to be packaged, carbonated, and shared. This chapter will guide you through the technical and creative aspects of bottling, from preparing bottles to choosing carbonation levels, and from ensuring safety to storing cider for the best possible flavor.

The Importance of Bottling Done Right

Bottling may seem straightforward, but it is one of the most critical stages in cidermaking. Done carelessly, it can lead to oxidation, contamination, or even dangerous bottle explosions. Done properly, it preserves the freshness, aroma, and flavor of your cider, while allowing you to achieve the style—still, lightly sparkling, or effervescent—that best matches your vision.

Think of bottling as the point where your work transitions from craft to experience. Every decision you make—carbonation level, bottle size, closure type—affects how your cider will be enjoyed.

Choosing Bottles and Closures

Glass Bottles

Most cidermakers use glass bottles because they preserve flavor, resist pressure, and showcase the cider beautifully. Brown or green bottles are recommended since they protect against light, which can degrade cider over time. Clear bottles look attractive but are better suited for ciders meant to be consumed quickly.

Bottle sizes include:

- **330 ml or 355 ml (standard beer bottle):** Perfect for personal servings.
- **500 ml:** A popular cider bottle size, balancing portability and sharing.
- **750 ml (wine or champagne style):** Ideal for special releases and sparkling ciders.

Closures

- **Crown Caps:** Common, affordable, and effective for sparkling ciders.
- **Swing Tops (Grolsch style):** Reusable and convenient for small batches.
- **Corks and Cages:** Traditional for champagne-style ciders, especially those requiring high carbonation.
- **Screw Caps:** Suitable for still ciders, though less common among artisans.

Carbonation: Still or Sparkling?

Carbonation is one of the most defining features of cider. Some styles shine as still, while others demand bubbles to enhance their crispness and aroma.

Still Cider

Still cider contains little to no carbonation, resembling wine in character. It emphasizes depth of flavor, tannins, and subtle aromas. To achieve still cider, you simply bottle without priming sugar and ensure fermentation is complete.

Sparkling Cider

Sparkling cider ranges from lightly fizzy to fully effervescent. Bubbles bring vibrancy, enhance aromas, and create a lively drinking experience. There are three main ways to carbonate cider:

1. **Bottle Conditioning (Natural Carbonation):**
    - Involves adding a small amount of sugar (priming sugar) before bottling.
    - The residual yeast consumes the sugar, producing CO_2 that

carbonates the cider naturally.

- Requires careful calculation to avoid over-carbonation and "bottle bombs."

2. **Force Carbonation (Kegging):**
    - The cider is placed in a pressurized keg and infused with CO_2.
    - Allows precise control over carbonation levels.
    - Bottles can then be filled from the keg using counter-pressure fillers.

3. **Traditional Method (Champagne Style):**
    - Involves secondary fermentation in the bottle with sugar and yeast.
    - Produces fine, persistent bubbles.
    - Requires disgorging and corking for a professional finish.

Calculating Priming Sugar

If you opt for bottle conditioning, accurate measurement is crucial. Too little sugar will result in flat cider; too much could create dangerous pressure.

A general rule:

- For moderate carbonation (similar to beer), add **4–6 grams of sugar per liter**.
- For champagne-like effervescence, add **7–9 grams per liter**.

Corn sugar (dextrose) is commonly used, though honey, maple syrup, or concentrated apple juice can add character. Always dissolve your sugar in boiling water, cool it, and mix evenly into your cider before bottling.

The Bottling Process: Step by Step

1. **Sanitize everything.** Bottles, caps, siphons, and any equipment must be sterilized to prevent contamination.
2. **Rack the cider.** Transfer the cider from the fermenter to a bottling bucket, leaving behind sediment.
3. **Add priming sugar (if carbonating).** Mix thoroughly but gently to avoid oxidation.
4. **Fill the bottles.** Use a bottling wand for

consistency, leaving about 1 inch (2.5 cm) of headspace.

5. **Seal bottles.** Cap, cork, or swing-top securely.
6. **Condition.** Store bottles at room temperature for 2–3 weeks to allow carbonation.
7. **Test a bottle.** Chill, open, and taste. If carbonation is satisfactory, move bottles to cold storage.

Storage and Aging

Once bottled, proper storage ensures that your cider remains stable and delicious.

- **Temperature:** Store bottles in a cool, dark place (ideally 50–55°F / 10–13°C).
- **Position:** Bottles sealed with corks should be laid horizontally; capped bottles should be stored upright.
- **Light:** Protect from sunlight and fluorescent light, which can damage flavor.
- **Duration:** Most ciders can be enjoyed within 6–12 months, though higher-tannin or barrel-aged ciders may develop complexity over several years.

Safety Considerations

- Always use bottles rated for carbonation. Reusing thin glass bottles (like those from some commercial soft drinks) can be dangerous.
- If in doubt, under-carbonate rather than over-carbonate.
- Label your bottles with dates and styles for easy tracking.

Bringing It All Together

Bottling is both technical and rewarding. It is the stage where your cider transforms from a fermenting liquid into a finished product—clear, bright, and ready for sharing. Each decision—whether you choose a still, sparkling, or champagne-style cider—shapes how your creation will be experienced. With careful technique, patience, and attention to detail, bottling becomes the bridge between cidermaker and cider drinker, sealing not just the liquid in the bottle, but the story of your work and the orchard it came from.

CHAPTER 9

Troubleshooting and Fine-Tuning

Even the most experienced cidermakers encounter challenges. Fermentation is a living process influenced by countless variables—temperature, yeast behavior, apple variety, and even air quality. While most batches progress smoothly, sometimes things don't go as planned. Off-flavors, stalled fermentations, and cloudiness can frustrate beginners and veterans alike. The good news is that most problems can be prevented, identified, and corrected with the right knowledge.

This chapter equips you with the tools to troubleshoot issues, rescue cider when possible, and fine-tune your technique so each batch improves on the last.

Common Problems and Their Causes

1. Stuck Fermentation

A “stuck” fermentation occurs when yeast stops converting sugar into alcohol before the cider is fully fermented.

Possible causes:

- Temperature too low or too high.
- Nutrient deficiency in the juice.
- Yeast strain exhausted or stressed.
- Excessive sulfite levels.

Solutions:

- Gently warm the fermenter to the yeast's ideal temperature range.
- Add yeast nutrient or energizer.
- Re-pitch with a fresh, active yeast starter.

- Ensure the must was not overdosed with sulfites before fermentation.

2. Cloudy Cider

Hazy cider is often harmless, but if you aim for crystal-clear presentation, it can be frustrating.

Possible causes:

- Suspended yeast or pectin.
- Insufficient racking or settling time.
- Infection introducing haze-causing microbes.

Solutions:

- Cold crash the cider by chilling it for several days to encourage sediment to fall.
- Use fining agents such as bentonite, gelatin, or pectic enzyme.
- Practice patience—many ciders clear naturally with time.

3. Unwanted Sweetness or Dryness

Cider may finish sweeter or drier than intended.

Possible causes of excess sweetness:

- Fermentation ended too early (stuck).
- Yeast strain incapable of fermenting all sugars.

Possible causes of excess dryness:

- Yeast consumed more sugar than expected.
- Highly fermentable juice composition.

Solutions:

- To retain sweetness, stabilize cider with sulfites and sorbate, then back-sweeten with sugar, juice, or honey.

- For overly dry cider, balance flavor with blending or back-sweetening after stabilization.

4. Off-Flavors

Unpleasant aromas or tastes can develop at different stages.

- **Sulfur (rotten egg smell):** Often from stressed yeast; aerate early in fermentation, provide nutrients, or rack cider.
- **Acetone (nail polish remover):** Indicates bacterial contamination;

prevent with strict sanitation.

- **Vinegar smell/taste:** Caused by acetobacter in the presence of oxygen; protect cider from air exposure.
- **Mousy or medicinal flavors:** Often from wild bacteria; proper cleaning and sulfite management help avoid them.

5. Bottle Bombs

Over-carbonation can cause bottles to burst, posing safety risks.

Causes:

- Bottling before fermentation is complete.
- Miscalculated priming sugar.
- Residual wild yeast continuing fermentation.

Prevention:

- Always confirm fermentation has finished (gravity readings stable for 2–3 days).
- Measure priming sugar precisely.
- Store conditioned bottles in a safe area until carbonation stabilizes.

Prevention Is Better Than Cure

The best troubleshooting strategy is prevention. By following good practices, many problems can be avoided entirely.

- **Sanitation:** Keep all equipment meticulously clean and sanitized.
- **Monitoring:** Use a hydrometer or refractometer to track fermentation progress.
- **Patience:** Rushing steps often leads to issues—give cider the time it needs.

- **Oxygen Management:** Limit air exposure after fermentation to prevent oxidation and bacterial spoilage.

Fine-Tuning Your Craft

Once you've mastered the basics and can troubleshoot confidently, the next step is refinement. Small adjustments elevate good cider to great cider.

Experiment with Yeast Strains

Different yeasts yield dramatically different results. For example:

- Champagne yeast: Clean, crisp, highly attenuative.
- English ale yeast: Fruity, with residual sweetness.
- Wild fermentation: Complex, rustic flavors.

Blending for Balance

Professional cidermakers often blend ciders from different apple varieties, fermentations, or even vintages. This creates depth and balance in flavor, tannin, and acidity. Try combining a dry cider with a sweeter one, or blending high-tannin cider with a lighter style.

Adjusting Acidity and Tannins

- **Acidity:** Too sharp? Blend with sweeter juice or mellow with aging. Too flat? Add malic acid or blend with sharper varieties.
- **Tannins:** Enhance structure by blending with bittersweet cider or using oak aging.

Aging Techniques

Time can transform cider, smoothing harsh edges and deepening complexity. Aging in stainless steel preserves freshness, while oak barrels add vanilla, spice, and tannic structure. Even a few months of bottle conditioning can noticeably improve balance.

When to Start Over

Not every batch can be saved. If cider develops strong vinegar, mold, or overwhelming chemical flavors, it may be best to cut your losses. Every "failed" batch is still a learning experience, teaching you how to refine your process.

The Cidermaker's Mindset

Troubleshooting isn't just about fixing problems; it's about developing resilience and curiosity. Each challenge deepens your understanding of fermentation and strengthens your connection to the craft. With time, you'll anticipate potential issues before they occur and confidently guide your cider toward the result you envision.

Remember: cidermaking is as much an art as it is a science. Problems are not setbacks—they are stepping stones toward mastery.

THE END

Printed in Dunstable, United Kingdom

73064604R00083